MARTIN XB-51 BY SCOTT LIBIS

INTRODUCTION

This book signals a radical departure from Ginter Books Naval Fighters Series. Previously, Army/Air Force planes were only covered when the aircraft was also in service with the Navy (Naval Fighters # 5 North American T-28 Trojan, NF# 8 Lockheed C-121 Constellation, NF #11 Grumman HU-16 Albatross, & NF #14 Convair T-29/C-131 series). Although hookless, there were many unusual post-war/early fifties Air Force developmental aircraft too. This is the story of arguably the most innovative and potentially the most promising of these development airplanes.

Anyone having photos or other information on this, or any other naval or marine aircraft, may submit them for possible inclusion in future issues. Any material submitted will become the property of NAVAL FIGHTERS unless prior arrangement is made. Individuals are responsible for security clearance of any material before submission.

ISBN 0-942612-00-0

Steve Ginter, 1754 Warfield Cir., Simi Valley, California, 93063

Above, XB-51 wooden mock-up on 26 February 1948 with the Army review board standing in front. Army officer (front row 3rd from right) was Captain Glenn Edwards. Tom Lanphier (2nd from left in front row, in civilian clothes) was the WWII fighter ace credited with shooting down Admiral Yamamoto's Betty bomber. He later went on to become the Assistant Secretary of the Air Force. The nose logo under Martin is a stylized XB-51 seen from 3/4 front and slightly below the horizon line. (Martin via Stan Piet)

BACKGROUND

Sleek swept wing aircraft burst onto the scene in this country in the late 1940s ushering in the next generation of aeronautical progress. Pleasing to the eye, these aircraft were exceptionally powerful and

Above, Martin's first jet bomber, the XB-48, utilized the main bicycle landing gear with outrigger wheels that Martin developed on its test-bed B-26, "The Middle River Stump Jumper", and on its XB-51. (via Tony Landis and Martin) Below, the B-45 went into production along with the B-47. (USAF) Below middle, Convair's elegant XB-46. (Convair) Bottom, Navy EB-47L. Boeing's B-47 was the real winner. (Berger)

agile. Newly emerging turbojet engine technology was another driving force propelling aircraft faster and higher. These swept wing turbojets easily outclassed anything piston powered and further inspired new design innovations that would leave their mark on aviation history.

In an attempt to keep up with the rapid advances being made in aerodynamics and powerplant technology, manufacturers were proposing and building new designs monthly. Some of these designs are now legendary, like North American's F-86 Sabre, and Boeing's B-47 Stratojet, while others never made it past the concept stage. Although many of these designs were never pressed into production, a few showed promise. It was these aircraft that became the testbeds that validated design features that would be so important to future aircraft. The Martin XB-51 was one of these testbeds.

MARTIN'S XB-48

The XB-48 was Martin's first jet bomber. Designed from Air Force specifications delineated in 1944, the XB-48 competed against North American's XB-45, Convair's XB-46, and Boeing's XB-47. With the competition set, Martin proceeded with the six-engine straight-wing XB-48. Housed on each slender wing was a huge engine nacelle containing three General Electric/Allison J-35 engines.

Another new design feature of the XB-48 was its bicycle landing gear, also used by Boeing's XB-47 entry. Bicycle landing gear, when retracted, was housed in the fuselage instead of the wing structure. Thus, wings could be designed thinner, eliminating drag penalties imposed by thicker wings. With a thinner wing, drag is reduced, and overall speed performance was increased. Martin engineers pioneered this bicycle landing gear on a heavily modified Martin B-26 that was referred to as the "Middle River Stump Jumper". Testing proved this landing gear configuration to be so maneuverable that competitor Boeing would later select it for both the B-47, and B-52 aircraft.

Advanced technology in aerodynamics captured from the Germans at the end of World War II eluded the Martin design team. This advantage would weigh heavily in Boeing's favor. Boeing's XB-47 featured a swept back wing, with pylon-mounted engine nacelles versus the XB-48's straight wing/buried wing nacelle design layout. In the end, the XB-48's top speed of 516 mph fell far short of the XB-47's 580 mph top speed and greater range. Martin lost that contract, and Boeing went on to produce one of the most memorable aircraft of all-time. These lessons learned from the XB-48 would greatly enhance Martin's next jet bomber.

At top, the bicycle development aircraft, "The Middle River Stump Jumper" during testing at Martin. (Martin via Stan Piet) Above, the A-26 was the aircraft that the XB-51 was to have replaced. The above A-26 was black with red trim and is seen in Korea. (via Gene Holmberg) Below, the Navy used A-26s designated JD-1s for utility work. JD-1D with Firebee under the wing on 19 August 1961. (Dustin Carter) Bottom, January 1947 drawing of the XA-45 which was re-designated XB-51 prior to its redesign. (Martin via Stan Piet)

MARTIN'S XA-45/XB-51

In 1946, another Army Air Force design competition solicited proposals for a new ground support aircraft to replace the aging Douglas piston engine A-26. The A-26 was a medium bomber, much smaller than the strategic bomber the XB-48 had been intended to replace. Martin's initial entry, model 234, was to be powered by two turboprops and two jet engines, with provisions for a crew of six. It would be a 70,000 pound airplane capable of carrying 8,000 pounds of ordnance 800 miles. Designated XA-45, Martin's model 234 won the competition. However, shortly after being awarded the contract the Army Air Force did away with the attack category, and Martin's model 234 was redesignated XB-51.

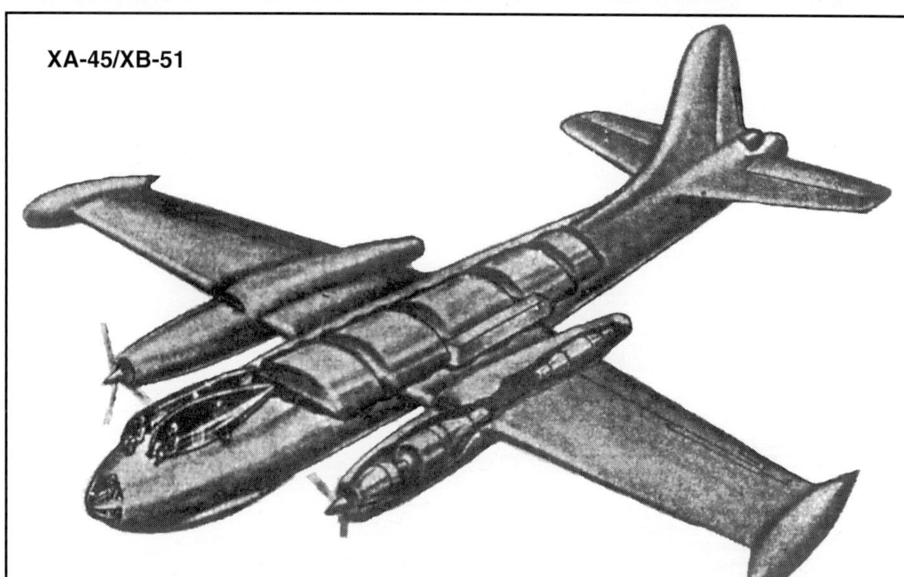

The A-26's replacement would be reclassified as a bomber. Emphasis on the new replacement would be on speed (640 mph top speed) with a range of 600 miles and a warload of 4,000 pounds. It was evident that Martin's original entry could not achieve the new speed criteria

In February 1947, Martin was ready with a new proposal. This radical new design was unlike anything ever seen before. It was a 52,000 pound airplane powered by three jet engines and included a variable incidence wing and tail, revolving bomb bay door and bicycle landing gear. Impressed with the new design, the

XB-51 PROPOSAL 15 JULY 1947

Below, artist sketches of the 1947 proposal which had the basic configuration and variable incidence wings of the prototype XB-51. The main difference was the lower tail-planes and heavily blended vertical tail.

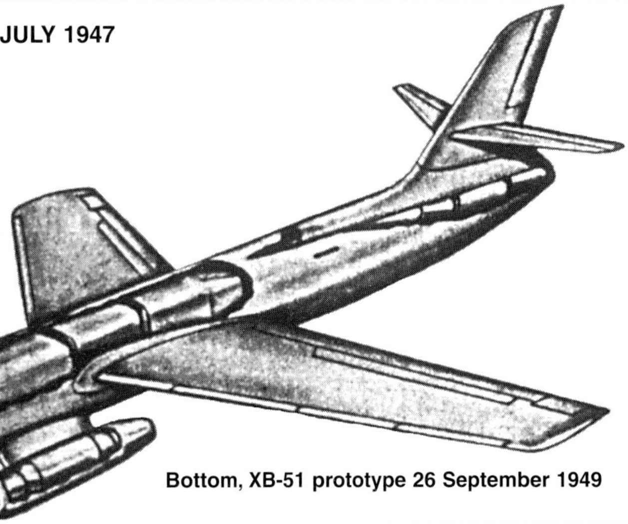

Bottom, XB-51 prototype 26 September 1949

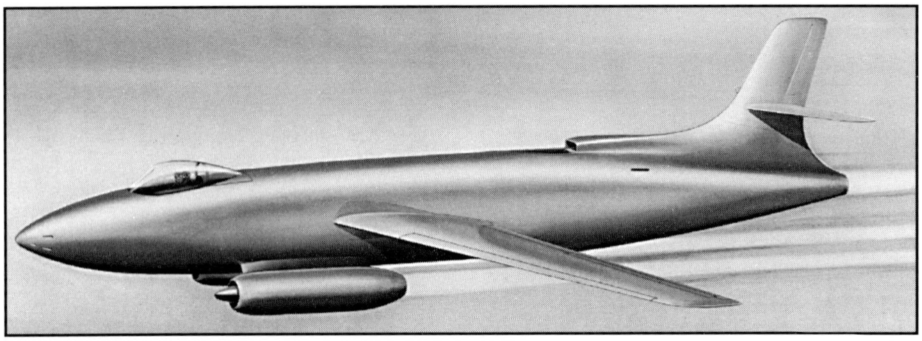

Air Force ordered two prototypes that would carry the XB-51 designation.

Starting out as a fixed price contract, Martin received $9,417,107 to produce two aircraft. This fee included associated wind tunnel models, mockups, special tools, spare parts, drawings, technical data and armament reports.

A wooden mockup of the XB-51 was completed and inspected by an Air Force Contractor review board on 24 February 1948. The Air Force assigned a young and promising test pilot named Glenn Edwards to the XB-51 program. Captain Edwards was also assigned as a test pilot on Northrop's YB-49 Flying Wing, then being tested at Muroc Field. Just two months after the XB-51's mockup review, Edwards lost his life while testing the YB-49. The giant flying wing was just north of the town of Mojave when it suffered a catastrophic structural failure in flight. An entry was later discovered in Edwards diary noting his visit to the Martin plant where he remarked that the XB-51 "promises to be a pretty good airplane." Had he survived, Edwards certainly would have been a key player in the testing of the XB-51. Muroc was later renamed Edwards Air Force Base in his honor.

Nicknamed the "Flying Cigar" because of the elongated fuselage, the XB-51 was powered by three General Electric J-47 turbojets. This configuration enabled it to outrun the F-80 and F-84 fighters that were assigned to fly chase on test hops.

The fuselage was configured with bicycle type landing gear balanced by a pair of wingtip outrigger wheels. Each of the main landing gear truck assemblies swiveled to permit cross wind landings. Positioned on either side of the fuselage, aft of the wings, were speed brake panels. When actuated, these speed brakes added to the XB-51's ability to bleed energy (slow down). And, for the first time in history, a single-point ground refueling capability was given to an airplane. This capability allowed ground maintenance personnel to fill every tank on the aircraft from one fueling point.

The #1 and #3 engines were contained in individual nacelles that were pylon mounted under the forward fuselage. These engines were positioned as close as possible to the aircraft's centerline to reduce any adverse effects of power loss on either engine. The #2 engine was housed in the aft fuselage, and used a rotating inlet at the base of the vertical stabilizer for engine air induction. This inlet could be opened for engine operation, or rotated closed during engine shut-down for increased inflight cruise fuel economy. Power from two up-and-running engines was more than adequate for in-flight operations. Four JATO bottles rated at 1,000 pounds of thrust each could be added to assist in maximum weight takeoffs.

Swept back at 35°, both wings and tail surfaces took advantage of the new advances made in aerodynamics. Sweeping the surfaces back, lessened the drag forces applied, allowing for higher speeds. The horizontal stabilizer placed at the tip of the vertical fin made it one of the first T-tails seen in this country. Turbulence generated by the fuselage was the determining factor in going from a mid-tail to a T-tail design. Another concern would be the blanketing effect that the variable incidence wing might have on the horizontal tail surface with the wing in the full up position.

The wings, although short like Martin's B-26, were surprisingly thin considering the complexity engineered into them. Without a doubt, this wing was one of the most ambitious engineering efforts to date. Housed within it were: leading edge anti-ice heating, automatic slats, full-span slotted flaps and an advanced spoiler-aileron. It's interesting to note that the only function the undersized ailerons served was to provide feel feedback to the pilot, whereas the bulk of the lateral control was provided by the spoilers. These spoilers gave the aircraft extraordinary roll control.

Above, artist drawings of the final T-tail configuration with wing tanks (top) and wing mounted Sparrow missiles (above). (Martin via Stan Piet) Below, prototype 6685 roll-out. Good view of the black wing and fuselage walk areas and T-tail upper surfaces. The upper fuselage intake door for the number 2 engine is rotated closed. Aircraft was nicknamed the "Flying Cigar" because of its elongated fuselage and abbreviated wings. (via Scott Libis)

SHIP #1 6685 ROLLOUT, 4 SEPTEMBER 1949

ROLLOUT

At left top, ship number one is towed out onto the "B" building ramp. The left side of the fuselage was not painted with the XB-51 designation. (Martin via Stan Piet) At left middle and bottom, right side view of 6685 showing the MARTIN and XB-51 on the forward fuselage painted in red and bordered by white. The square block script on the XB-51 was repainted in italics to give the impression of speed prior to the first flight. The wings were swept-back at a 35 degree angle. The aircraft had an extremely small canopy for the pilot and a porthole on each side of the fuselage for the Shoran operator, whose compartment was located aft of and lower than the pilot's station. The tail-mounted engine intake door is rotated closed. (Martin via Stan Piet)

At left top, the XB-51 seen from behind shows off many of the aircraft's innovations. The high T-tail, rear mounted jet engine with its intake duct faired into the base of the vertical fin, the variable incidence downward drooping wings, and bicycle gear with outrigger wheels. (Martin via Stan Piet) At left, head-on view. The original T-tail was later modified with a bullet fairing at the junction of the vertical fin and horizontal tail-planes. A small, flat black anti-glare panel was painted in front of the pilot's windscreen. (Martin via Stan Piet) Below, 3/4 front view with the flaps lowered 30 degrees. The World War Two camouflage is still evident on the Martin factory buildings in the background. (Martin via Stan Piet)

XB-51

Wing variable incidence control allowed the whole wing to change angles, from three to seven degrees of positive incidence. Simply put, incidence is defined as: the angle between the wing cord line and the airplanes longitudinal (pitch) axis. It was this feature that enabled the XB-51 to land at such low speeds. Another advantage wing incidence offered is that when in use, it would lower the aircraft's nose, dramatically improving pilot visibility for landing approach. Vought's F-8 Crusader later successfully used this lift concept for carrier operations. In addition to the incidence control system, the aircraft was designed with 6° of wing anhedral which in appearance made the wings sag at the tips. The horizontal stabilizer also used a variable incidence controller.

In what was thought to be the answer to high level, high speed bombing stability, Martin engineers Werner Buchal and Albert Woolens fitted the XB-51 with a revolutionary new rotating bomb-bay door. Internally-stored bombs or rockets would be mounted to the door itself, then prior to reaching the target the door would be rotated 180 degrees open and locked into position. This advance meant that the aircraft did not have to slow down to open the bomb-bay doors, as the gap between the fuselage and the door was only momentary. Conventional bomb-bay doors remain open for the entire bomb run, generating considerable buffeting, which in turn reduces stability and bombing accuracy. By maintaining maximum speed the aircraft's vulnerability to enemy fighters and anti-aircraft weapons would be greatly minimized. Another advantage was

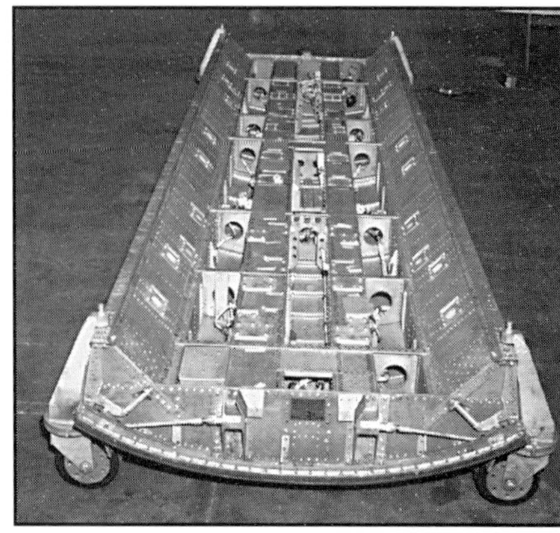

Above, revolutionary rotary bomb bay door which could be pre-loaded on the ground and hoisted within minutes as shown in the test rig below. Bottom, the bomb bay door is fitted with flight test instrumentation and is rotated partially open for access. (Martin)

At far left top, the XB-51 blasts off with a high performance takeoff. Note the extended leading edge slats. (Martin via Stan Piet) At left top, wing trailing edge showing the small ailerons at the end of the wings, the upper wing spoilers, the lowered flaps, and the bleed air wingtip exhaust vents. The wing-to-fuselage end plate that normally covers the fuselage slot in which the variable incidence wing moves is absent. Note the heavy wrinkling of the aft fuselage skin and the two open auxillary intake doors below the engine. (Martin via Stan Piet) At far left middle, close-up of the upper wing with slats on the leading edge and the spoiler visible on the aft edge. (Martin via Stan Piet) At left middle, small extended speed brake with a good view of the aft main landing gear doors. The rotary bomb bay is partially open and degree markers are painted on the fuselage aft of the wings so that chase planes can verify wing and flap settings. The two small bumps on the top of the fuselage next to the engine intake are the fuel vent housings (Martin via Stan Piet) At left bottom, structural tests on the wing control surfaces of ship #1 on 19 May 1950. The wingtip bleed air exhaust vents show up well as does the forward outrigger gear door. (Martin via Stan Piet)

that spare rotating bomb-bays could be pre-loaded off the aircraft, and be ready for quicker mission turnaround times (much like a speed-loader for a revolver).

A pilot and a SHORAN operator were the only crew that the cramped interior could accommodate. Cockpit pressurization was provided and offered a small measure of crew comfort. The pilot enjoyed fighter-like visibility with a small bubble canopy that seemed even smaller in proportion to the long, narrow fuselage. The second crew member operated the short-range navigation and bombing systems just below and to the aft of the pilot station. Ejection seats were provided for emergency egress for both crew members.

PILOT'S RIGHT - HAND CONSOLE AND CIRCUIT BREAKERS

At left top, view of pilot's instrument panel through the small bubble canopy. At left bottom, right side of the pilot's forward cockpit shows the cavernous cockpit area below the canopy sill line. The thumb paddles on the control yoke were for activating the JATO bottles. At right, pilot's right-hand console contains the engine starting panel and the radio and lighting panels. The three large levers on the right are the control surface locks. Below, the pilot's right-hand circuit breaker panel is located just to the right of the control panel locks. At right middle, looking down at the shelf or upper portion of the pilot's right-hand console. The engine starting controls are at left with the lighting controls at the right. Bottom right, pilot's right-hand console vertical or lower portion contains the radio and interphone controls. All photos are of the second XB-51 as indicated by the taped-on note above the radio which states RADIO CALL 6686. (Martin via Stan Piet)

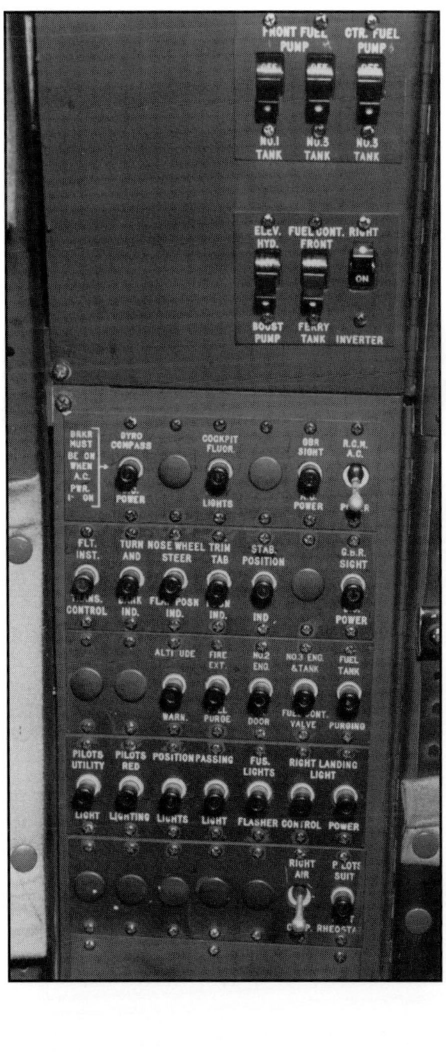

PILOT'S LEFT - HAND CONSOLE AND CIRCUIT BREAKERS

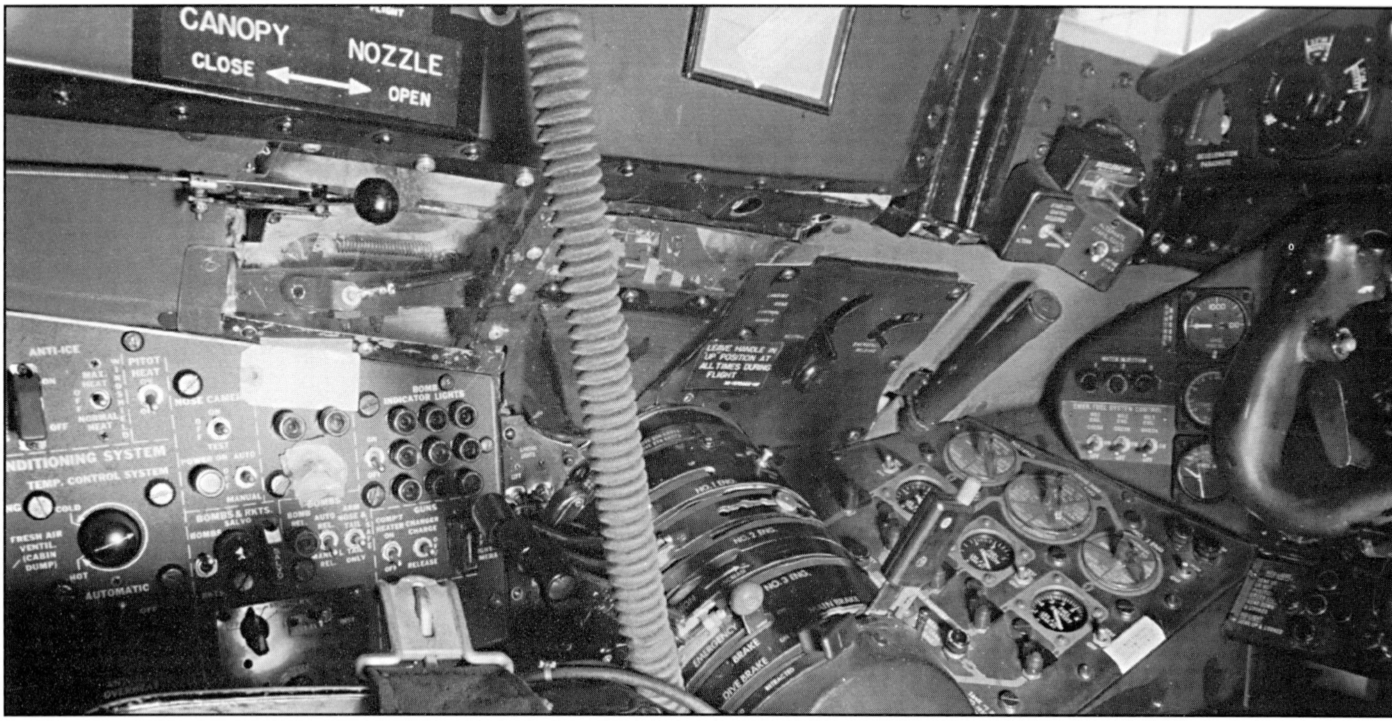

Above, pilot's left-hand console, with the fuel management panel just forward of the throttle quadrants. The lower indicator lights are for the two 375 gallon ferry tanks when installed. Bottom, left-hand pilot's console aft of the throttles which contain circuit breakers, air conditioning, bombing & gunnery panel, dive brake & flap controls. (Martin via Stan Piet)

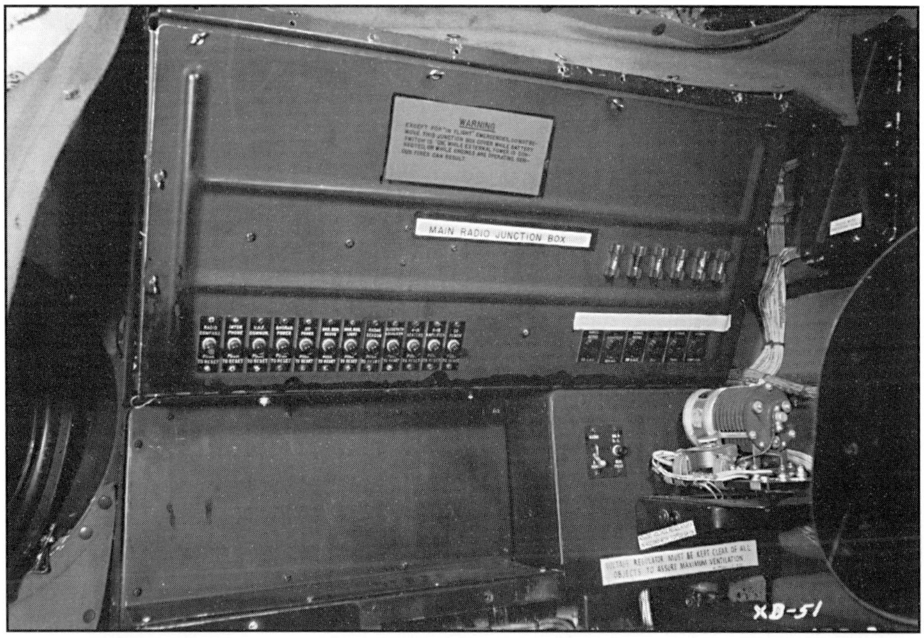

Top right, main radio junction box in the upper right Shoran compartment. The seat headrest is at the far right and the fuselage side window is at the far left. Above, right side camera controls and oxygen regulator were located directly below the main radio junction box and above the radio and oxygen panels at right. Below, vibration testing to the Shoran operator's ejection seat. At lower right, left-hand side of the Shoran operator's compartment showing the circuit breakers, lighting controls and flying suit heater control. (Martin via Stan Piet)

EIGHT 20MM CANNON GUN NOSE INSTALLATION

A novel feature made its appearance on the second XB-51 prototype. It was a removable nose section allowing the aircraft to change mission configuration. There were three different nose packages that could be quickly changed by ground maintenance crews. The first was a bombing nose which contained the equipment necessary for high accuracy bombing. The second, a reconnaissance nose, housing an array of cameras for reconnaissance purposes. And finally a gun nose which contained eight 20mm cannons, carrying 120 rounds of ammunition each (see the appendix page twenty-seven for additional aircraft design and performance specifications).

The XB-51 was not only innovative, but it was built for ease of maintenance, as these photos help illustrate. At left, the gun nose viewed from above through two large access doors. Another door is visible in the forward upper door. Below, the right side of the gun nose with the single large maintenance door removed. Three of the 20mm cannons are visible. (Martin via Stan Piet)

THE J47-GE-13 ENGINE

The XB-51 owed much of its nimble performance to three General Electric J47 axial flow turbojet engines. The engines' general arrangement consisted of a 12-stage compressor feeding an 8-can combuster section. Driving the compressor was a single-stage turbine flowing back into a two flap variable exhaust nozzle. The flaps, mounted externally, would open and close like eyelids increasing, or decreasing, the exhaust cross-section. 5,000 pounds was the thrust rating on this engine and gave the XB-51 15,000 pounds of thrust at military setting. Additional thrust augmentation for takeoff was available by using the onboard water/alcohol injection system.

Above right, J47 engine on the engine test stand at Martin on 11 July 1950, with a P4M-1 Mercator behind it and hangars 4, 5, & 6 in the background. At right, complete podded engine test cell for the #1 and #3 engines undergoing testing at Martin. The wing section and nacelle on the platform were in use since 1940 for the Mars prototype. At right bottom, the variable exhaust nozzle and mechanism. The podded engines were designed for quick removal by unbolting the lower half and then using portable bomb hoists to lower the engine. Below, the variable nozzle on the #2 engine. (Martin via Stan Piet)

Above, the number two XB-51 poses for a PR shot with many of its potential weapons. Shown are: 5" HVAR rockets, 500 lb. bombs, 750 lb. bombs, 1,000 lb. bombs, and a 4,000 lb. bomb. Below and at right, cluster bombs fitted in the XB-51's rotary bomb bay door ready for hoisting into place. Bottom, aerial flares loaded into the XB-51's rotary bomb bay. The door is rotated slightly to the right. Normally an additional bomb would be carried where the camera blister is located. (Martin)

Accessories driven by the engines were: starter-generator, tachometer, fuel regulator, fuel pump and lube and scavenge pumps. Other Air Force aircraft using the J47 included the F-86, B-36 and B-47. Several models of the J47 were produced with a total production run of over 35,000 units.

WEAPONS

Initial flight testing uncovered a new obstacle to high-speed bombing. Never before had bombs been dropped at such high-speeds, and it was discovered that the bombs tended to tumble and were sometimes kicked back up into the bottom of the airplane. To resolve this problem, Martin engineering devised a pair of pneumatically actuated pistons and affixed them to each bomb shackle. Now the bomb would be ejected, rather than relying on gravity for positive separation.

Two bomb bay door configurations were built offering the Air Force greater flexibility in mixing ordnance. The "A" door was the flush fitting door most generally used. Photo flares, cluster bombs, 500/750lb bombs and 5" HVAR rockets were the intended weapons for this "A" door. Two additional hard-points were added to the outside of the door for carriage of two 2,000 lb general purpose bombs.

The "B" door was tailor-fitted around the dimensions of a single

Above and at right, 3-views of 500 lb. bombs loaded in the rotary bomb bay. Above view shows the two cameras used to photograph bomb drop tests. Below, the rotary door viewed from behind with 750 lb. bombs loaded. Bottom. 4-1,000 lb. bombs loaded in the XB-51. Bottom right, 6-750 lb. bombs loaded on the rotary bomb bay door are positioned for hoisting into the XB-51. Once the door is hoisted off the ground the four wheels were removed before the door was raised into place. (Martin via Stan Piet)

4,000 lb (LC) general purpose bomb. From the side, this door bulged down below the streamlined fuselage. There were no provisions for external hardpoints on this door as they would not clear the ground.

Among the wide variety of weapons intended for carriage on the XB-51 was the mysterious "special weapon". The term "special weapon" is the designation used when discussing classified weapons, or bluntly put, atomic bombs. The successful employment of two atomic bombs brought Japan to its knees during WWII and ended the most destructive conflict in the history of mankind. Exploiting this capability would indeed give the U.S. the upper hand in any future military actions. Consequently, Air Force and Navy planners immediately embarked on a campaign to ensure that every new non-fighter combat aircraft design would incorporate this new special weapons capability.

The "B" door appears to have the proper dimensions to conceal a Mk.-5 bomb. The Mk.-5 is a direct descendant of the "Fat Man" device dropped on Nagasaki, Japan, in 1945. With an overall length of 132 inches, the bomb

At top, 2-2,000 lb. bombs mounted externally. At left, 5" HVAR rockets fitted to the rotary door. Bottom left, the bulging "B" door which was built for a 4,000 lb. bomb or a Mk.-5 "special weapon". Below, a 4,000 lb. bomb nestled in the XB-51's bomb bay. (Martin)

FLY - OFF, XB-51 VERSUS THE ENGLISH ELECTRIC CANBERRA (B-57)

had a 43 inch diameter and weighed 3,175 pounds. Total destructive force was put at 81 kilotons. Design refinements led to a Mk.-7 which was also intended for carriage on the XB-51.

THE FLY - OFF

In 1951, with the Korean conflict in full swing, the Air Force set up a review board at Andrews AFB to find a replacement for the Douglas A-26. The candidates were: North American's AJ-1 Savage, Canada's CF-100, English Electric's Canberra, and Martin's XB-51. These aircraft were also evaluated for their suitability as night attack aircraft, with extended range and loiter time considered crucial. Although slower than the XB-51, the Canberra's lower wing loading gave it a range and low altitude maneuvering advantage. The Canberra could loiter for two-and-a-half hours over a target 900 miles from base, compared to only one hour 400 miles from base for the XB-51. Even so, COL Albert Boyd, chief of the Flight Test Branch at Edwards AFB recommended the XB-51 to the Air Council over the Canberra. The review board, however, saw matters differently and declared the Canberra the winner, recommending that an order for 300 aircraft be placed.

Now known as the B-57 Canberra, Martin was licensed to build the Canberra at the Maryland plant, due in part to the backlog at the English Electric factory. 403 B-57s would ultimately be produced by Martin in several different versions.

The fate of the XB-51 was sealed when the Canberra won the contract. However, the Air Force retained both prototypes at Edwards for flight testing of new and unproven technology.

At right, the AJ-1 Savage was considered along with the Canadian CF-100 as replacements for the A-26. At right middle, Canberra Mk.B2 (WD-940) was the second pattern aircraft received by Martin. At right, production Martin B-57 with tandem cockpits.

XB-51 SHIP #1 (46-685) SIGNIFICANT EVENTS

Above, O. E. 'Pat' Tibbs poses in front of the XB-51 prior to the first flight. Bottom, Tibbs starts the taxi for the XB-51s maiden flight on 27 October 1949. The ship is fitted with a single JATO bottle on each side of the fuselage. (Martin via Stan Piet)

Above, on 19 October 1949 a tri-service inspection and taxi demonstration was conducted by Martin's chief test pilot O. E. 'Pat' Tibbs. It is unknown what purpose the cord exiting the Shoran operators window was used for. The auxiliary intake doors for the #2 engine were open on the after-fuselage during taxi. (Martin)

On 27 October 1949, using both water injection and JATO, Ship #1 took to the air for the first time with company test pilot Pat Tibbs at the controls. During this first flight the aircraft proceeded to 10,000 feet where a series of clean and dirty stalls were made. The dive brakes were tested, and then another climb was made to 20,000 feet. Thirty-four minutes after take-off, the airplane landed at nearby NAS Patuxent River. Landing rollout was only 3,000 feet due in part to the deceleration chute. Phase I flight testing had commenced.

EARLY JATO TAKEOFF RUNS

JATO takeoffs were common during the early phases of the test program and like most early jet engines, the J47 was a real "smoker" at military power settings, which added greatly to the smoke and turbulence of the JATO motors. (Martin)

XB-51 LANDING AND DRAG CHUTE

On 28 December 1949, the first of several landing mishaps occurred. The aft main landing gear retracted after a ground run of 2,000 feet, allowing the aircraft to settle on its aft section. Cause of the uncommanded retraction was believed to be the sudden application of vertical and forward loads. This could have caused the gear to unlock and collapse. Replacement of the aft landing gear doors cost $21,000.

At left, drag chute being installed in the tail of the XB-51. The chute door opened on the left side of the aircraft for deployment of the chute and for maintenance. Below, RT-80 shadows the XB-51 down to landing. Once on the ground the chute was popped to slow down the aircraft. Bottom, the aft main gear failed on 28 December 1949. (Martin via Stan Piet)

"**Phase 1 airworthiness; flight test conducted by the contractor to demonstrate the aircraft's ability to fly safely.**"

After 45 test flights, Phase 1 flight testing was completed on Ship #1. Phase 1 revealed surprisingly few deficiencies in the XB-51 design. Those requiring changes were:

A) Addition of a fairing at the tail intersection to eliminate airplane vibration and elevator roughness at high Mach.

B) Hydraulic and mechanical changes to the main landing gear.

C) Redesign of the wing flap control and actuating system because of the increase in flap loads with spoilers deflected.

D) Change in lateral control system from feel aileron and spoiler to all-spoiler. 100% boost configuration, including an electrically operated trim spoiler because of feel aileron reversal at high speed and low altitude.

The Air Force's Air Material Command initiated Phase II flight test activities on 31 March 1950, at Martin's Baltimore facility.

"**Phase II contractor compliance: flight tests, in which the Air Force pilots fly the aircraft to determine if it meets the performance guarantees.**"

Ship #1 had been in Phase II for a little over a month when, after 16

At right top, with the engine removed and the #2 engine doors open, the damage to the doors from the aft main gear failure can be seen. At right middle, ground down after fuselage lip caused by the gear failure. The XB-51 had a large area aft of the bomb bay and beneath the #2 engine intake duct which would house most of the aircraft's electronic equipment and some camera equipment. During the early test flights a photo panel was installed against the aft main gear bulkhead as seen here. At right, improved view of the photo panel on the bulkhead and the radio and Shoran equipment installed on the right side of the fuselage. The intake duct can be seen in the top of the photo. (Martin via Stan Piet)

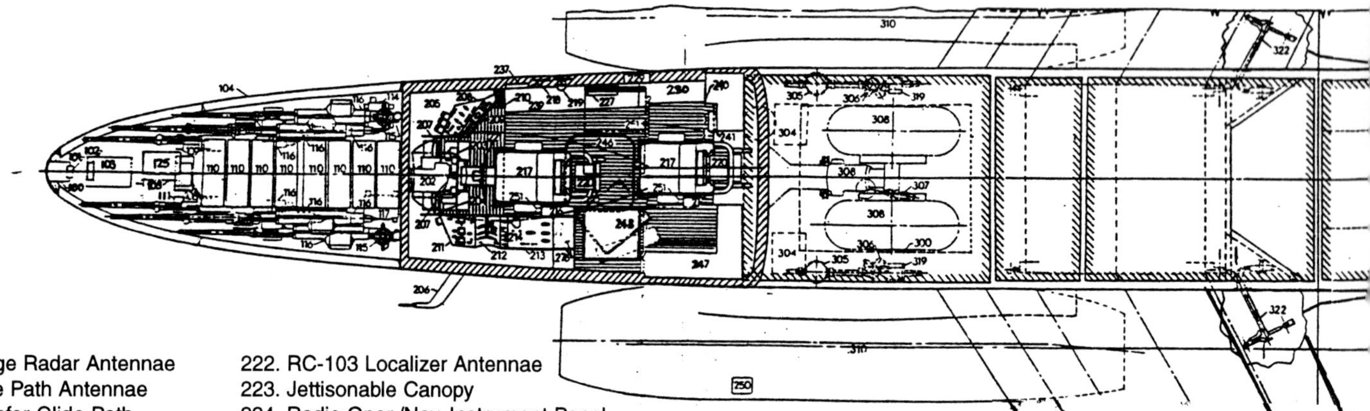

100. Range Radar Antennae
101. Glide Path Antennae
102. Transfer-Glide Path
103. Radar PGE RF Unit
104. Nose Compartment Doors
105. Radar Range Unit
106. Camera window
107. Nose Cap
108. Shutter Trip Unit
109. Camera Station
110. Ammo Cans
111. Box
112. Shoran Rec. Antennae
113. Nose Strike Camera
114. Computer
115. Compressor
116. Guns
117. Amp
118. Gun Barrels
119. Radar Beacon Rec. Ant.
120. Radar Beacon Trans. Ant.
121. Valve
123. Forward Bulkheads
124. Shell Case Compartment
125. A-1-B
200. Lower Switch Panel
201. Armor-Glass Windshield
202. A-1-B GBR. Sighting Head
203. Gun Sight Panel
204. Pilot's Main Instrument Panel
205. Pilot's Right Hand Console
206. Pitot Tube
207 Rudder Pedals
208 Electrical Panel
209. Radio Panel
210 Control Lock Oper. Levels
211. Fuel Panels
212. Engine Controls
213. Armament Panel
214. Emergency Valves
215. Intervalometer (Bombing)
216. Map Case
217. Ejection Seat
218. Crash Axe
220. Stabilizer Chute
221. Control Column
222. RC-103 Localizer Antennae
223. Jettisonable Canopy
224. Radio Oper./Nav. Instrument Panel
225. Heating & Air Conditioning Equip.
226. Radio Oper./Nav. Table
227. Signal Flare Cont. Type-A6
228. T-10 Pyrotechnic Projector
229. Radio Oper./Nav. Window
230. Radio & Radar Junction Box
231. Intervalometers (Camera)
232. Camera & Radio Compass Control
233. Oxygen Regulator
234. Radar Countermeasure Control
235. Gyrosyn Amplifier
236. Gyro Control
237. RH Elect. Circuit Breaker Panel
238. Map Case
239. Controls Guard
240. Recirculating Air Intake
241. First Aid Kit
242. Filler Cap-Hyd Reservoir
243. Voltage Reg. & RH Elect. Center
244. Radio Wire Pressure Box
246. Cabin Entrance Door
247. LH Electrical Center
248.ID-17/APN-3 Shoran Indicator
249. CM-3/APN-3 Shoran
250. External Power Receptacle
251. Relief Tube
252. APW-11 Radar BN Rcvr-Trans.
253. AM-26/AIC Interphone AMP
254. AT-134/ARN Marker Beacon Ant.
255. Equipment Access Door
256. R-122/ARN-12 Marker Beacon Rec.
257. R-89 B/ARN-5B Receiver
258. RC-103 Localizer Receiver
259. APW-11 Dynamotor
260. Inverter
261. Battery
262. Emergency Hyd. Reservoir
263. Engine Flak Plate
264. Accumulator
265. Motor & Hyd. Pump
266. Step
267. Entrance Door
268. AM-142/AIC Mix. AMP Interphone
270. Generator Control Switches
271. Chart Case
272. Hydraulic Relief Valve
273. Pressure Regulator
274. Radio Oper./Nav. Escape Hatch
275. Radio Oper/Nav. Radio Console
276. LH Elec. Circuit Breaker Panel
277. Jet Orifice Control
278. Air Bottle
300. Tank Access Door
301. Tank Filler Door
302. Radio Wire Conduit
303. Hydraulic Reservoir
304. Oil Tank
305. Fuel Tank
306. Fuel Selector Valve
307. Fuel Shut-Off Valve
308. Front Main Landing Gear
309. Front Gear Doors
310. External Jet Engines
311. Wing Incidence Drive Mech.
312. Door Engine Air Intake
313. Boundary Layer Bleed Exit
314. Fuel Purge System
315. Fuel Selector Valve
316. TR Gear
317. Rear Main Landing Gear
318. Rear Gear Doors
320. Bomb Bay Door
321. Landing Light
322. Wing Flap Control
323. Rear Engine Intake
324. Boundary Layer Bleed Ent.
325. Fuel Tank Purge Station
326. Reverse Flow/Windmilling Door
327. Bomb Hoist Door
328. Controls Guard
329. Elect. Wire Duct
330. Single Point Fuel Door

1/72 SCALE

331. Bomb Door Operating Mech.
332. Bomb Bay Tank Filler Door
400. AS/313 A/ARN-6 Loop Ant.
401. Radio Compass Sense Ant.
402. Rear Engine Air Duct
403. Auxiliary Air Duct
404. Fuel Gage Amplifiers
405. BC-966 IFF Trans.
406. SA-3A IFF Inertia Switch
407. R101/ARN6 Radio Compass Rec.
408. Type 333A-1 Rad. CCVIP. Servo Amp
409. J-68/ARC-3 VHF Power Junction Box
410. T-67/ARC-3 VHF Trans.
411. R-77A/ARC-3 VHF Radio Receiver
412. BC-767 IFF Indicator Box
413. Vertical Camera Station
414. T-11/APN-3 Shoran Trans.

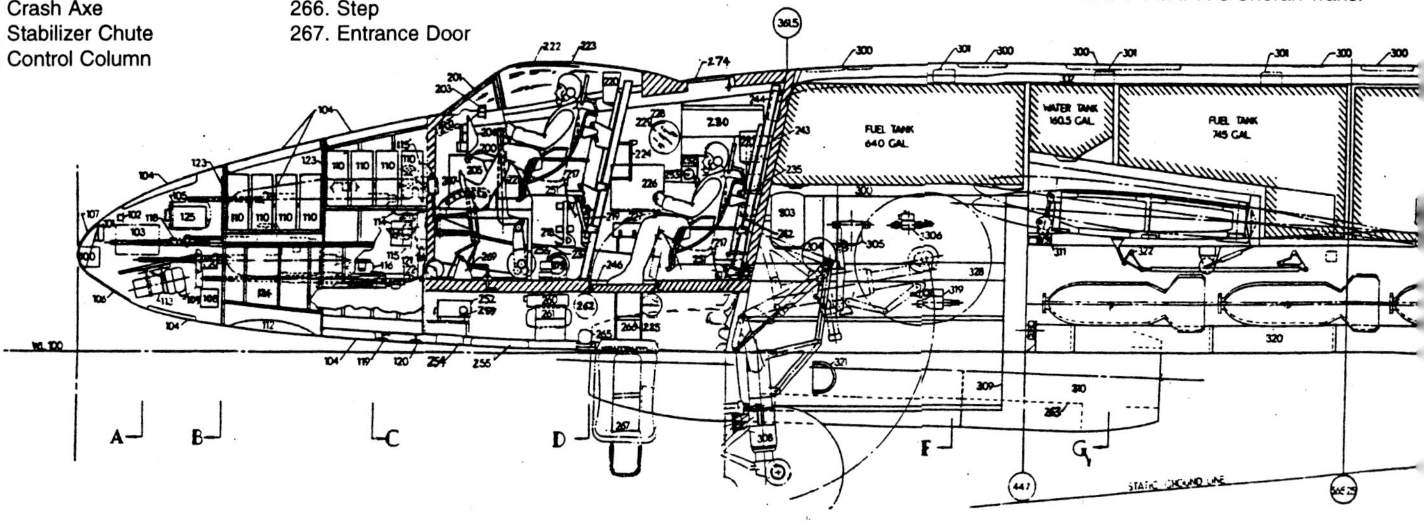

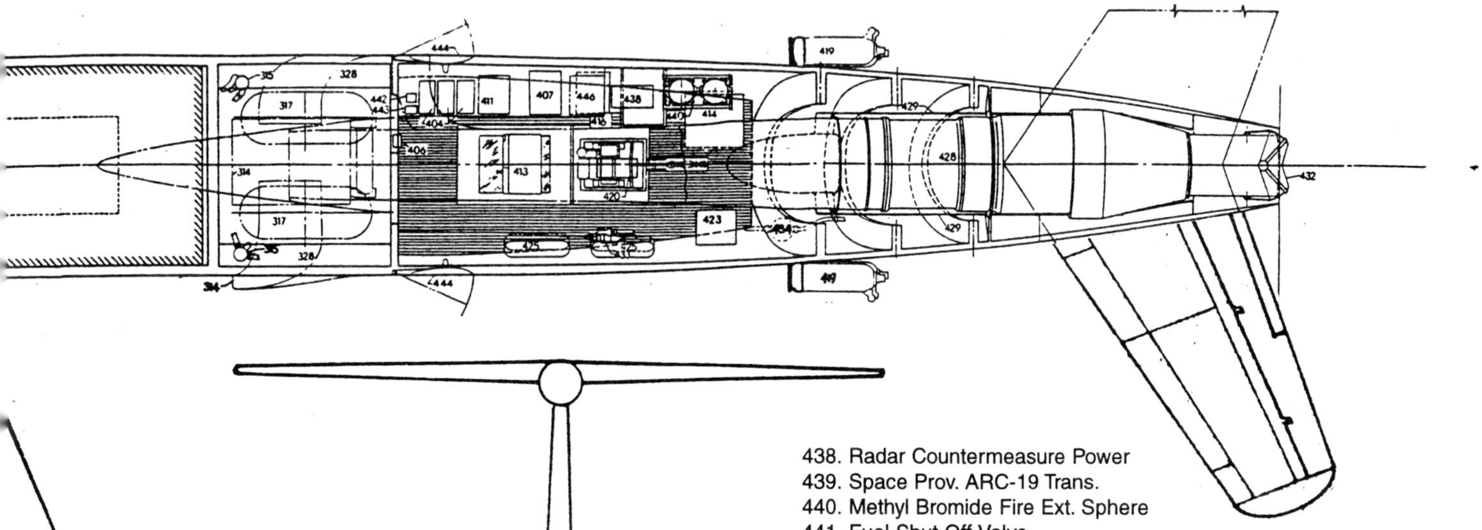

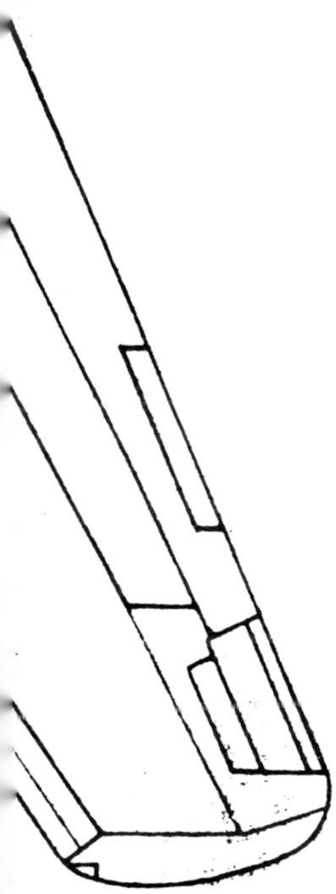

438. Radar Countermeasure Power
439. Space Prov. ARC-19 Trans.
440. Methyl Bromide Fire Ext. Sphere
441. Fuel Shut-Off Valve
442. Fire Detector Relay Box
443. No. 2 Engine Jet Nozzle Control Box
444. Dive Flap
445. Removable Door
446. Electronic Control-Purge System
447. Electronic Countermeasure Modulator

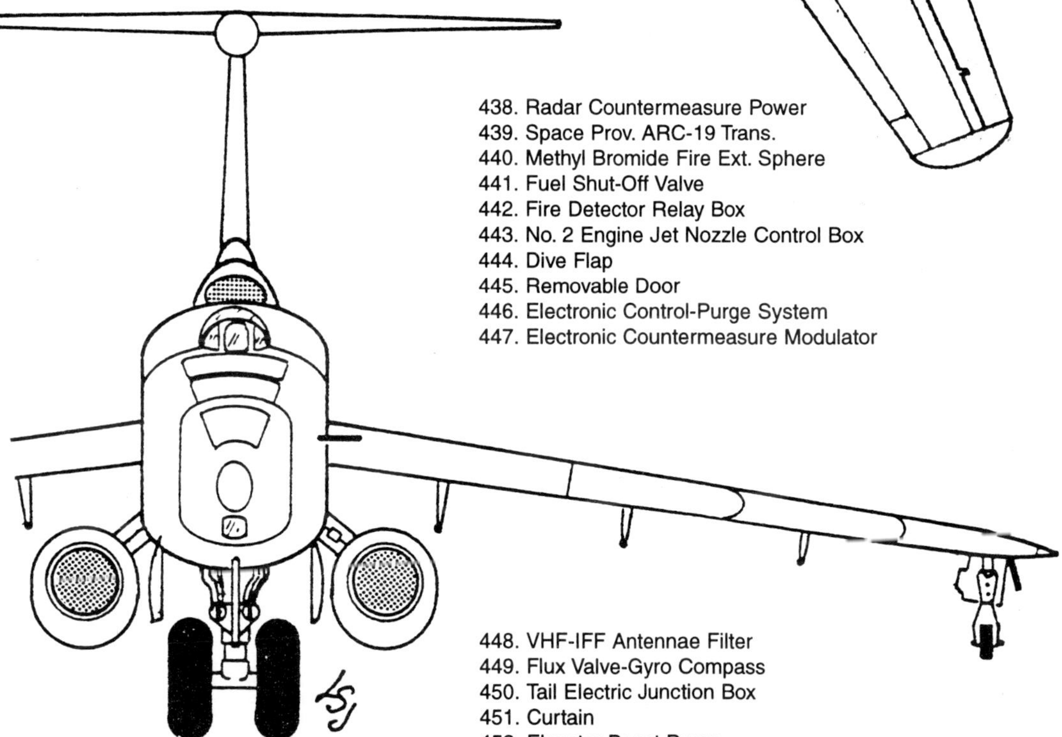

448. VHF-IFF Antennae Filter
449. Flux Valve-Gyro Compass
450. Tail Electric Junction Box
451. Curtain
452. Elevator Boost Pump
453. Elevator Boost Hydraulic Reservoir

415. Camera Vacumn Pump
416. Shutter Trip Unit
417. Camera Window
418. Door Curtain
419. JATO Bottle
420. Rear Camera Station
421. Rear Camera Door
422. Oil Tank Filler Door
423. Oil Tank
424. Shoran Trans. Ant.
425. Oxygen Bottles
426. VHF Antennae
427. Engine Shroud
428. Rear Jet Engine
429. Engine Flak Plate
430. Hinged Cowling
431. Exhaust Nozzle Actuating Motor
432. Variable Area Exhaust Nozzle
433. Water Pump
434. Fuel Filter
435. Camera Doors & Operating Gear
436. Radar Countermeasure Trans.
437. Radar Countermeasure Ant.

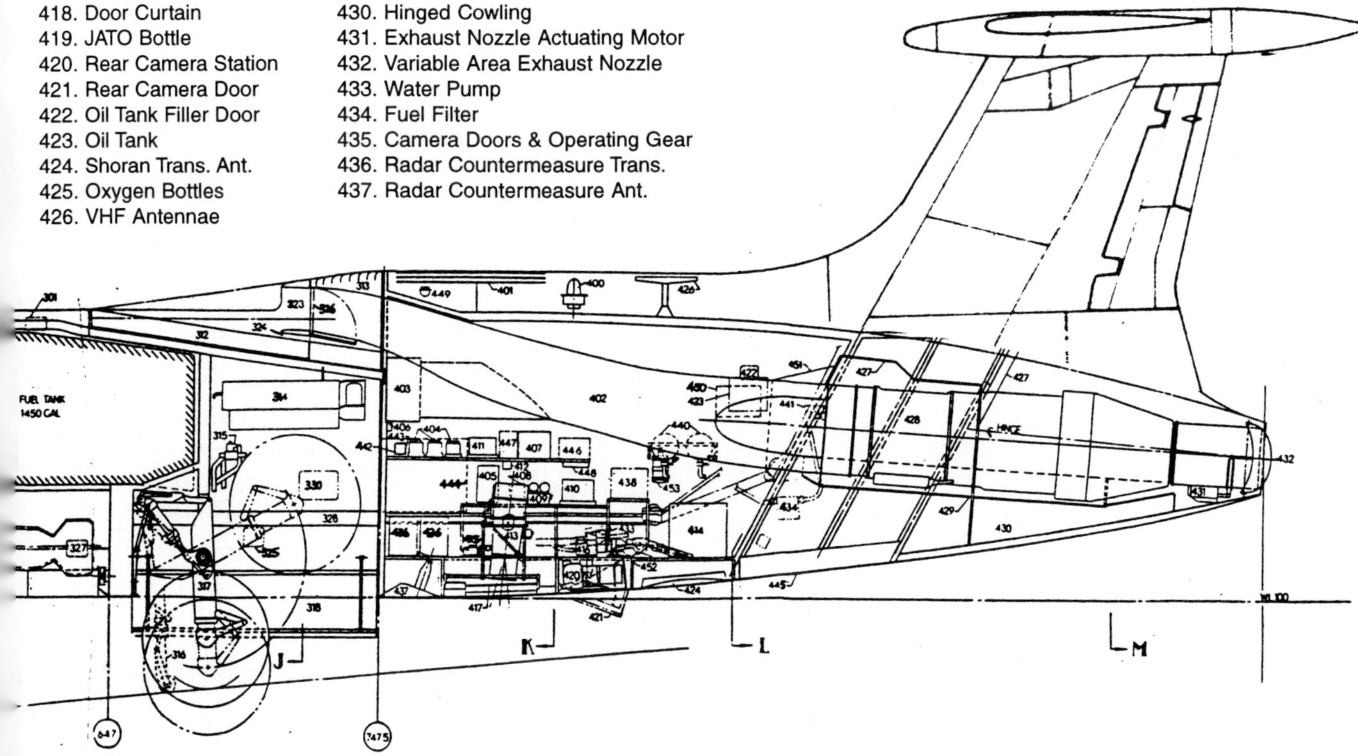

GENERAL LAYOUT 1/144 SCALE

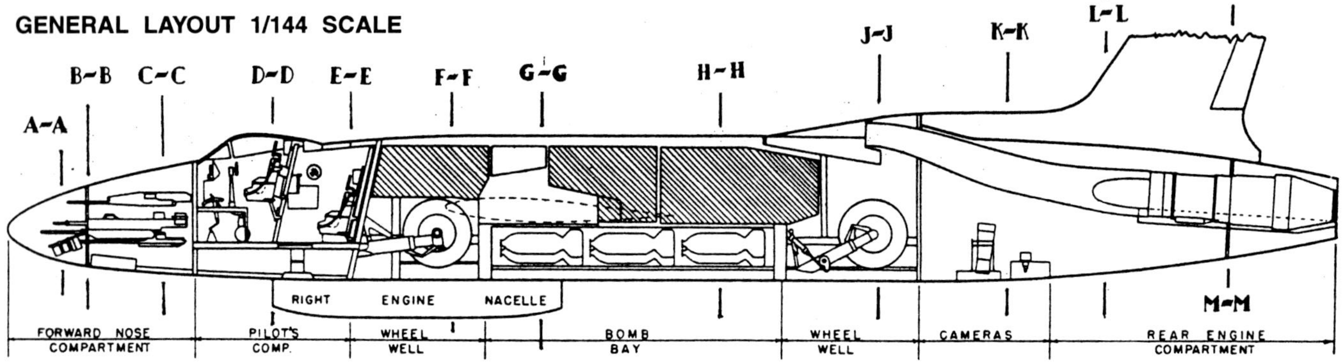

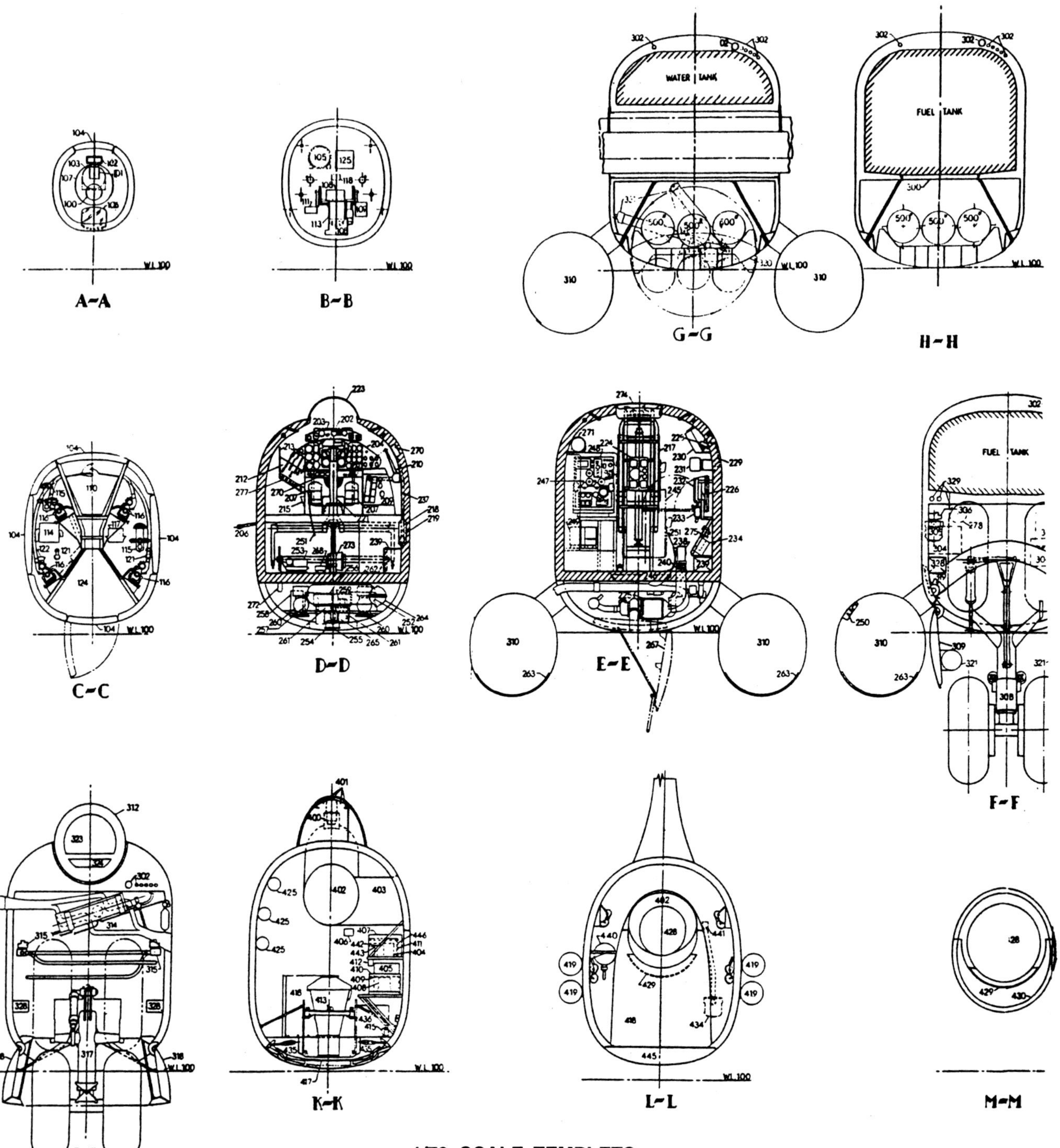

1/72 SCALE TEMPLETS

WEIGHTS

Loading	Lb	L.F.
Empty	29,584	
Basic	54,648	3.56
Design	53,000	3.67
Combat	*41,457	3.67
Max T.O.	**62,452	2.0
Max Land	+57,067	1.6

*Basic Mission
**Limited by strength
+ Limited to landing without external bombs and water

BOMBS

No.	Size	Type
1	*4000	L.C.
2	2000(int)	G.P.
2	2000(ext)	G.P.
4	1600(int)	A.P.
2	1600(ext)	A.P.
4	1000	G.P.
9	500	G.P.

Max Bomb Load: ..... 10,400 lb
Max Bomb Size: .... 4000 lb
*In special door

GUNS

No.	Cal	Rds ea.	Loc.
8	20mm	160	Nose

ROCKETS

No.	Size	Type	Loc
8	5''	HVAR	Bomb bay

Note:

Rockets can be carried in bomb bay in lieu of internal bombs.

POWER PLANT

No. & Model .... (3)J47-GE-7
Mfr. .... General Electric
Engine Spec. No. E581-A
Type .... Axial Flow
Length .... 144''
Dia. .... 37''
Weight (dry) .... 2525 lb
Augmentation: ..... Fluid Injection

FUEL

Location	No.Tanks	Gal
Fuselage*	3	2835
Bomb bay	2	700
*Self-sealing	Total	3535

Spec. AN-F-58
Grade JP-3
Water/Alcohol(gal) 160.5

OIL

Capacity (gal) 16.2
Spec. .... AAF 3606

DIMENSIONS

Span .... 53.1'
Length .... 85.1'
Height .... 17.3'
Tread (outriggers) .... 49.4'

ELECTRONICS

VHF Command AN/ARC-3
Radio Compass AN/ARN-6
Localizer RC-103
Glide Path AN/ARN-5B
Marker Beacon AN/ARN-12
Radar Beacon AN/APW-11
Shoran AN/APN-3
RCM AN/APT-16
IFF SCR-695B
Interphone .... AAF Combat

ENGINE RATINGS

S.L. Static	LB - RPM
T.O:	*5820 - 7950
Mil:	5000 - 7950
Nor:	4250 - 7370

*Fluid injection

Jato Units

4 x 1000 lb (14 sec duration)

Loading and Performance - Typical Mission

CONDITIONS			BASIC MISSION	ATTACK MISSIONS RADIUS	ATTACK MISSIONS RADIUS	HIGH ALTITUDE MISSIONS RADIUS	HIGH ALTITUDE MISSIONS RADIUS	HIGH ALTITUDE MISSIONS RADIUS	FERRY RANGE
			I	II	III	IV	V	VI	VII
TAKE-OFF WEIGHT	⑦	(lb)	55,923	55,923	54,137	55,923	58,238	62,452	56,662
Fuel & Oil		(gal)	2835/12	2835/12	2835/12 ⑨	2835/12	2835/12	2835/12	3535/16.2
Military Load		(lb)	4000	4000	1120	4000	6384	10,494	None
Total Ammunition		(rds/cal)	1280/20mm	1280/20mm	1280/20mm	1280/20mm	1280/20mm	1280/20mm	1280/20mm
Wing Loading		(lb/sq ft)	101.7	101.7	98.4	101.7	105.9	113.6	103.1
Stall Speed -(power off)		(kn)	133	133	131	133	136	140	134
TAKE-OFF DISTANCE SL									
Ground Run (no wind)	④	(ft)	4340	4340	4010	4340	4790	5965	4480
To Clear 50ft Obst	④	(ft)	5590	5590	5190	5590	6200	7835	5790
CLIMB FROM SL									
Rate Of Climb at SL	② ⑧	(fpm)	5100	5100	5340	5100	4890	3960	5030
Time To 10,000 Feet	②	(min)	2.2	2.2	2.1	2.2	2.4	3.0	2.3
Time To 20,000 Feet	②	(min)	5.4	5.4	5.3	5.4	5.7	6.3	5.6
Service Ceiling (100 f.p.m.)	②	(ft)	32,900	32,900	33,700	32,900	31,600	27,200	32,500
COMBAT RANGE		(n.mi)	934	—	—	1044	—	—	1401
COMBAT RADIUS		(n.mi)	378	331	332	516	502	446	—
Avg. Cruising Speed	⑥	(kn)	463	440	440	466	466	463	467
Total Mission Time	⑥	(hr)	1.82	1.68	1.70	2.40	2.33	2.12	3.08
Cruising Altitude (1000 ft)	⑥	(ft)	32.8 to 43.2	10.0	10.0	32.9 to 43.2	31.7 to 43.3	27.6 to 43.2	32.8 to 43.0
COMBAT WEIGHT		(lb)	41,457	41,958	43,118	40,747	40,571	40,289	—
Combat Altitude		(ft)	S.L.	10,000	10,000	35,600	34,300	29,700	—
SPEED									
Max Speed (combat alt)	②	(kn)	560	552	552	503	506	519	—
Max Speed At S.L. Ft	②	(kn)	560	560	560	560	560	560	—
CLIMB									
Rate Of Climb (combat alt)	②	(fpm)	6980	5270	5080	1360	1600	2350	—
Rate Of Climb At SL	②	(fpm)	6980	6900	6650	7130	7160	7190	—
CEILING									
Combat Ceiling	②	(ft)	38,900	38,600	38,100	39,200	39,300	39,400	—
Service Ceiling	②	(ft)	40,500	40,300	39,800	40,900	41,000	41,100	—
Service Ceiling	③	(ft)	39,900	39,700	39,100	40,200	40,300	40,400	—
LANDING WEIGHT SL	⑥	(lb)	34,267	34,267	35,361	34,267	34,198	34,302	34,962
Ground Roll	④	(ft)	2355	2355	2425	2355	2350	2360	2400
From 50' Obst	④	(ft)	3240	3240	3325	3240	3235	3245	3295

NOTES
① Take-off power
② Max power
③ Normal power
④ Take-off and landing distances are obtainable at sea level using normal technique. For airport planning add 25% to distances shown
⑤ Detailed descriptions of the Radius & Range missions are given on page 6 & 7.
⑥ Radius mission if Radius is shown
⑦ All T.O. weights include 1275 lb water/alcohol for T.O. (1 minute)
⑧ T.O. gross wt less T.O. water and fuel for 5 min. at N.R.P.
⑨ 8 x 5'' HVAR rockets

CONDITIONS:
(a) Performance Basis: Estimated data
(b) In computing Radius and Range, specific fuel consumptions have been increased 5% to allow for variation of fuel flow in service aircraft.
(c) Performance is based on powers shown on page 7.

EARLY TEST FLIGHT CONFIGURATION

Above, right side view of the first XB-51 during flight number 13. Of the four major changes dictated by the 45 test flights conducted during Phase 1 flight testing, only one affected the physical appearance of the airplane. This was the development of a bullet fairing at the juncture of the "T" tail to eliminate vibration at high Mach number. These photos show the original tail configuration. Below, overhead view showing wing, fuselage and tail walk areas bordered in black. (Martin via Stan Piet)

"T" TAIL AND FIN MODIFICATIONS

Above, during Phase 1 testing a thin metal fin fence was fitted in an attempt to control vibration of the tail. At right bottom, Ship #1 on 3-10-50 with its new bullet fairing at the "T" juncture. Martin ground crew top to bottom: William Knoll, Charles Liberto, Boyd Shiner, Ken Seippel, and Frank Ballentine. Below left, on 11-16-51, the fin fence was damaged when tried again.

SHIP NUMBER ONE TAIL FAIRING MODIFICATION

test flights had been completed, a disturbing lateral trim condition demanded immediate attention. Ship #1 was returned to Martin for a "fix", as well as installation of an M-1 canopy remover for emergency ejection. Martin struggled through the summer to find a fix for the ailing lateral trim, and an additional 17 flights were made without any improvement. Next, the flaps received reinforced control linkages in hopes of improving control. Another five flights were made with the reinforced controls, apparently with favorable results, and the Air Force resumed flight testing in mid-September.

27 September 1950 saw another landing mishap. An Air Force pilot made a hard landing necessitating the repair/replacement of the aft engine doors. On 10 November 1950, the Air Force completed Phase II after 41 flights and 44 hours of flight time. The Air Force reported that the airplane's performance was excellent, and equaled or exceeded the contract's specifications. Although a "Dutch Roll" tendency still existed, it was acceptable with the incorporation of the modifications for wing incidence and flap unloading.

At top left, 685 makes a high-speed pass over Martin's runway. At left above, with the #2 engine doors partially open 685 sits on the ramp on 2-28-50 with the new tail fairing. At left, front view of the tail fairing. Below, 685 is being prepped for a test flight as 686 takes off in the background. (Martin)

Above, 685 after shedding its forward landing gear on 2-28-52 from landing short at Wright-Patterson AFB. The crew blew the canopy, and even though the aircraft appears relatively undamaged it took $408,000 to fix. (from Seth Caperton via Peter Merlin)

"Phase III design refinement: conducted by the contractor, using the same aircraft as in phases I and II, to overcome deficiencies noted during the previous tests."

This next phase was referred to as the "production prototype program". Martin once again assumed test responsibilities that included: studies, investigations, design evaluation, design changes and flight testing. This effort was to aid in the evolution for a production supersonic bomber. Tests were performed on wing and tail airloads, nacelle and nacelle pylon airloads, high-speed vibration, and forced tail vibration. 39 flights later, ship #1 was returned to the Air Force.

Now-retired Air Force Inspector Howard Keller recalls an incident involving Ship #1 on 7 May 1951. "We had just pulled the chocks for Colonel John Calopy, the pilot for that particular hop, when the #3 engine caught fire. Fortunately, he hadn't gone too far and we had a fire bottle, so the fire was extinguished quickly." The following excerpt from the accident report tells the rest. Investigation revealed that the fuel tanks had been topped off on Friday evening and that fuel had expanded over the weekend and run into the tank vent lines. Every time the pilot braked the airplane during taxiing, some of the fuel spilled from the vent lines onto the hot #3 engine. The resulting fire damaged the nacelle, nacelle pylon, engine, and related wiring. Keller goes on to comment, "Placement of the #1 and #2 fuel tank vents would have to be re-examined if the 51' were to go into production."

Plans now called for ship #1 to join Ship #2 in its high-speed bombing tests at Muroc, California (Edwards). However, before its delivery to Muroc, Ship #1 was badly damaged in yet another landing accident at Wright-Patterson AFB, Ohio. During the landing approach, the Air Force pilot flared too late. The airplane made a hard landing short of the approach end of the runway, tearing off the front landing gear wheels, and damaging the lower-foreword

Above, two of the Air Force pilots involved in the program, Captains Robert Mapp and Lee Horn. Below, 685 arrives at Martin by truck on 6-16-52. Tape covered the XB-51 script on the forward fuselage. (Martin via Stan Piet)

Above, the forward landing gear seen from the front with landing lights mounted on the landing gear doors and on the landing gear's scissors. Below left, interior of the landing gear well looking forward with the backs of the landing lights on either side. At right top, Decelostat brake installation on the aft side of the nose gear on Ship number 2 on 28 February 1951. At right bottom, rear main landing gear looking aft at the Decelostat brake installation. Note simplistic gear door mechanism.

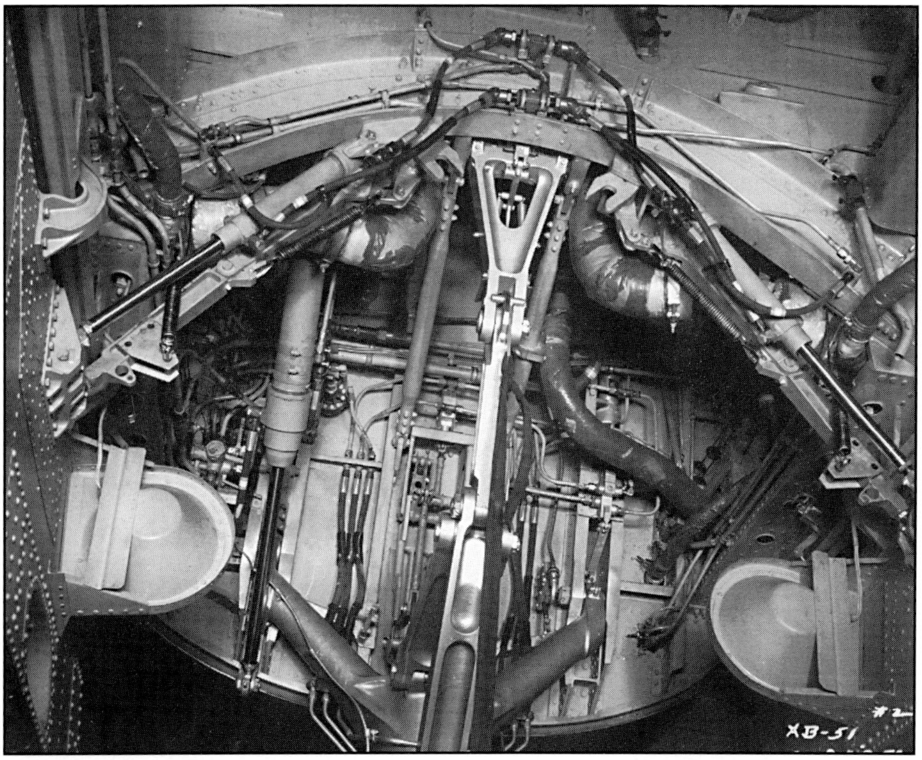

section of the airframe. The damage was so extensive that the aircraft had to be disassembled and transported by truck back to Martin for repairs. The cost of these repairs reached $408,000.

With repairs complete, Ship #1 was ferried to Edwards AFB to complete high-speed bombing tests. It was also during this time that the XB-51 starred in the movie "Toward The Unknown". The airplane was dubbed the XF-120 Gilbert fighter for this 1955 movie. Test pilot Lincoln Bond, played by actor William Holden, set out to discover the cause of a stability control problem. Doing the stunt flying for actor William Holden was Air Force Major Pete Everest. Everest, a consummate Test Pilot, put on an impressive demonstration with the XB-51, performing crisp snap rolls and impressive climbouts. Major Everest, also known as "The Fastest Man Alive", did the X-2 flights in the movie too, and served as one of the picture's main technical advisors. This movie was filmed on location at Edwards AFB, and many other early X-planes are documented throughout the movie.

At right top, actors William Holden (right) as Maj. Lincoln Bond, and Lloyd Nolan as Gen. Banner. (via Tony Landis) At right, Paul Fix as Gen. Shelby (left), Bartlett Robinson as Senator Black and Lloyd Nolan watch a tanker demonstration. Below, William Holden in the cockpit of the X-2. At right bottom, Murray Hamilton as Maj. Lee (left) looks on as Lloyd Nolan briefs Bill Holden. (via Scott Bloom)

SHIP #1 OPERATIONS AT EDWARDS AFB (MUROC)

At left top, dramatic JATO takeoff from Edwards AFB with a fleet of F-86s and a B-47 in the background. At left middle, 685 on the ramp at Edwards. A makeshift canopy cover protects the cockpit from the desert heat. The #2 engine intake door is rotated shut and the lower engine and aft fuselage access doors are unlatched and hanging free. (USAF via Scott Libis) At left bottom, 685 banks low over Muroc dry lake as it returns from a test flight. (USAF via Tony Landis)

The final blow was dealt to the program when Ship #1 was lost on 25 March 1956, while on a ferry flight from Edwards AFB to Eglin AFB. During a stop at El Paso Municipal Airport, eye witnesses watched the ill-fated takeoff. The airplane appeared to accelerate slowly to approximately 7,000 or 7,500 feet down the runway. The airplane then lifted off the runway in a slight nose-high position. The gear was retracted immediately, but the airplane failed to climb. Takeoff was continued, and the aircraft passed over the right side of the runway about 500 feet from the end. Shortly after passing the end of the runway, the airplane started to settle. This continued for about 1,000 feet, at which time the airplane struck the boundary fence. The airplane continued to slide through the terrain for about 750 feet from the boundary fence, disintegrating piece-by-piece until it stopped and burned.

At right, late in the test program, 685 sported red da-glo lower wing and forward fuselage hi-viz markings. On the exterior of the bomb bay door are two bomb pylons which were used for 2,000lb. bombs. In addition to performance testing, extensive wea-pons testing was conducted on the XB-51 at Edwards AFB. (Martin via Stan Piet)

The airplane's crew chief was killed and the pilot, Major James Rudolph, was severely burned. He died from his injuries a few days later. Cause of the accident was attributed to premature take-off rotation leading to a stall. At the time of the accident Ship #1 had logged a total of 432 flight hours over approximately 453 contractor/Air Force test flights.

Above, 685 in flight over Southern California. Good view of the wing and fuselage walkways which were outlined in black and the Shoran operators window located aft of the pilot's bubble canopy. (Martin via Stan Piet)

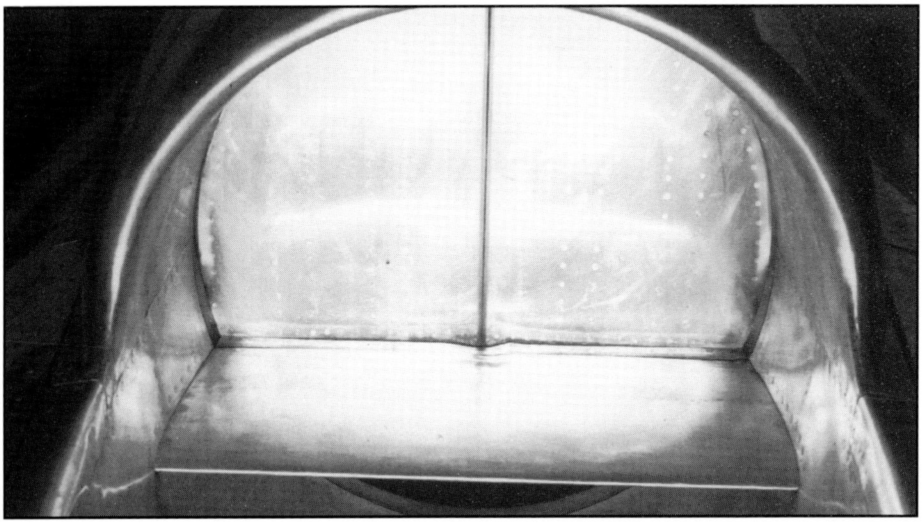

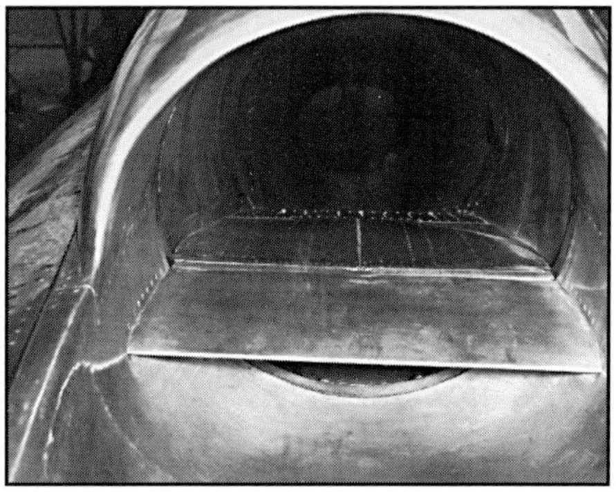

The XB-51 had an anti-reverse flow door installed in the #2 engine intake duct which was closed when the #2 engine was not being operated in flight. Upper left, door shut. Above, door opened by folding back into the duct. (Martin via Stan Piet) At left and below, 685 undergoing maintenance on the #2 engine at Edwards AFB. The extremely thick engine doors are braced in the full-open position. Below, bomb hoists were used to raise or lower the engine into position. The buzz number XB-685 was added to the rear fuselage shortly before the aircraft's demise. (via Wayne Morris) At right, fuel spill episode at Edwards AFB. Top right photo shows good detail of the outrigger gear. At right bottom, good view of the slats in the extended position. Note the ARDC insignia on the forward fuselage. (USAF via Scott Libis)

US AIR FORCE
6685
XB

XB-51
U.S. AIR FORC
XB-685

XB-51 SHIP #2 (46-686) SIGNIFICANT EVENTS

Phase 1 flight test began on 17 April 1950 when Ship #2 flew for the first time. Company pilot F. E. "Chris" Christofferson logged only 10 minutes due to a landing gear retraction problem. With the gear retraction problem solved, testing activities shifted to armament testing at the Edgewood Arsenal, Maryland. Upon landing at the arsenal to begin rocket testing, however, Ship #2 was damaged during the landing roll. Pilot landing technique was blamed for running the aircraft off the runway and shearing the forward main landing gear. After $96,000 in repairs, Ship #2 was returned to service.

Above left, the #2 XB-51 was rolled out from "C" building on 16 February 1950 with the original "T" tail installed. The bullet fairing was added to the "T" tail prior to its first flight. At left, the first taxi test was conducted on 16 April 1950. Below, JATO bottles are fitted on 686 the day of its first flight on 17 April 1950 with 685 in the background. (Martin via Stan Piet)

SHIP NUMBER TWO (46-686) FIRST FLIGHT 17 APRIL 1950

Above, the second XB-51 is prepped for its first flight on 17 April 1950. Compare the flap down and the flap up position between the two aircraft. Below, 686 makes its first take off on 17 April 1950. This flight only lasted ten minutes because because of a landing gear retraction problem. (Martin via Stan Piet)

By this time it was apparent that the XB-51 would not win a production contract, so emphasis was placed on developing the innovative rotary weapons bay door that would ultimately see service on the B-57 Canberra. The XB-51 was the perfect test bed for this high-speed, low drag component. With its snug fit, buffeting at high subsonic speeds was cut to a minimum providing a highly stable bombing platform.

By 29 January 1951, it had dropped 500, 2,000 and 4,000 pound bombs at 625 IAS and fired rockets in ripples and dropped in salvos at the

Above, Ship #2 taxis in to the ramp after a chute landing on 26 April 1950 after an early test hop. Below, close-up of 686's nose with huge gun access door and forward strike camera clearly visible. Bottom, 686 on 1 June 1950. Note the extremely long tow bar used on the XB-51. (Martin via Stan Piet)

same speed. Results of the testing were impressive with only three bombs tumbling out of a total of 56 dropped.

On 14 December 1951, testing moved to Edwards AFB where additional data was collected on bomb release characteristics of the XB-51 configuration. Another objective was to determine the highest possible speed that could be obtained in level flight. Testing was uneventful until March 21st. While conducting power-off stalls at 15,000 feet the right hand aft gear door opened in flight, causing

Above, 686 taxis out during weapon tests with externally mounted 2,000 lb. bombs on its bomb bay door on 10 July 1950. Below, with JATO lit, 686 starts its take off run with its external bomb load. Bottom, rare two-ship photo of the two XB-51s over Martin Field on 11 October 1950. (Martin via Stan Piet)

Below, in October 1950, 686 was tested with the aft bullet "T" tail removed. Note tufting applied to the tail for the test. (Martin via Stan Piet)

SHIP #2 OPERATIONS AT EDWARDS AFB

an excessive air pressure to surge in the aft section of the aircraft. This in turn dislodged a floating section of the #2 engine inlet duct and allowed fragments to be drawn into the #2 engine. The engine was shut-down and the aircraft landed safely.

JATO (jet assisted take off) entered a new phase with ship #2 and a new liquid fuel motor. Previously, these assisted take off devices burned a solid propellant which had about the same consistency as pencil erasers and once lit could not be turned off. Liquid fuel motors, however, were more stable and could be turned on or off. This new motor used a two-part fuel made up of Nitric Acid and JP-4 jet fuel. Major benefits of the new motor were: lighter reusable motors which reduced the take off roll of the 55,000 pound XB-51 from 5,200 feet to 4,175 feet.

Tragedy struck the XB-51 program when ship #2 was lost on 9 May 1952 while making low-altitude, high-speed flybys at Edwards AFB. On the fourth pass the pilot raised the nose and started an aileron roll. The roll appeared normal until the aircraft became inverted, at which point the nose dropped below the horizon. The aircraft then appeared to "dish out" of the roll, striking the ground in a left wing low, slightly nose-down attitude. The aircraft exploded and burned upon striking the ground, instantly killing the pilot, Major Neal Lathrop. Fortunately, the SHORAN station was unoccupied for this flight. Pilot error was ruled as the cause of the accident. At the time of the accident ship #2 had accumulated 151 flight hours over 193 contractor and Air Force test flights.

Top left, Edwards AFB gate in 1953. Sign reads: "Air Force Flight Test Center." (Martin) Above left, 686 banks over the lake bed and comes in for a landing. Note that the gunports are uncovered. At left, Edwards ground personnel service 686 with Nitric Acid which was used in the newly-developed liquid fuel JATO rockets. (USAF via Scott Libis)

Above, XB-51 parked next to a B-47 during the filming of "Towards The Unknown". (via Scott Bloom)

POLITICAL CONCERNS

Post-World War II led to a massive scaling back on defense spending, and a battle began to brew between the Navy and the newly formed Air Force, which became a separate service in 1947. This battle pitted the Navy's aircraft carriers against the Air Force's strategic bombing force, in particular Convair's massive B-36 bomber then on the drawing boards. The Peacemaker had tremendous range and was capable of striking European targets from bases in the continental United States. This placed the B-36 program in direct competition with the Navy's aircraft carriers for budget purposes.

The XB-51 may have become a victim of this debate when company founder Glenn L. Martin backed the Navy and lobbied for support. Although the Navy was Martin's biggest customer, some felt that Glenn Martin may have been overzealous in his stance on the "Bomber versus Carrier" debate. Martin was accused of circulating information critical of then Secretary of Defense Louis Johnson. Johnson, a former Convair executive, was alleged to have been biased against the Navy, ultimately leading to the cancellation of its supercarrier "USS United States".

This 1949 time frame coincides with the XB-51 design team's attempts to win a production contract. General Hoyt Vandenburg, Chief of Staff, and Secretary of the Air Force, Stuart Symington, allegedly vowed that "it would be a cold day in hell before the Air Force bought another airplane from Glenn Martin."

THE PILOTS

Far away from Washington's political bickering, on the west coast history was being made in the skies over Southern California's high desert. The Bell X-1 and its now famous pilot, Captain Charles Yeager, had broken the sound barrier. Yeager was the most visible member of the Mach 1 test team. Other members included:Major Robert Cardenas, the pilot of the B-29 mothership which launched Yeager and the X-1 into fame. Yeager was further backed-up by the Deputy Chief of Flight Test,

At right, 686 with open gun ports and canopy cover. This view of the nose illustrates the flat nose glass of the forward firing strike camera. 685 is being towed in the background. (Martin)

Colonel Fred Ascani, who participated in the selection of Yeager for the job and oversaw the X-1 and other ongoing test programs. And so, it was amid this hoopla that XB-51 ship #2 arrived at Muroc for phase II flight testing.

Flying 635.686 mph in an F-86E,

At left, Air Force test pilot Colonel Fred Ascani was a veteran test pilot, he retired as a Major General. (USAF via Scott Libis) Below, Air Force Captains Lee Horn and Robert Mapp (in helmets) were involved with the XB-51 at Martin. At far left is William Sparrow, Martin flight engineer. Martin pilot Frank Christofferson, who made the first flight in 686, is nearest the crew hatch. (Martin via Stan Piet)

Air Force Colonel Fred Ascani set a new 100 km closed course speed record at the 1951 National Air Races. A decorated WWII veteran, Ascani became a test pilot, flying more than 50 experimental prototype and research aircraft. One of these 50 plus aircraft stands out, deserving special recognition according to the now retired Major General. "The XB-51 was without a peer in its day, it was way ahead of anything else. In fact, it was comparable to fighters in the performance that it could sustain. On the deck, at full speed the XB-51 would have been extremely difficult to intercept in a fighter".

"There were no unpleasant, or unfavorable, control conditions on the deck at high dynamic "Q" when you were cruising at Mach .92/.93. The control characteristics were very simple, easy to fly. It's important to note that the proper attitude had to be established for landing. You had to hold to a proper speed. If you were too fast, you would end up touching the nose wheels first. If you were too slow, you would touch the rear wheels first and fall forward on the nose gear, possibly wiping it out. So you strove mightily on the final approach to establish the proper attitude, so that when you touched down, you had both trucks of wheels touching down simultaneously".

"Takeoff performance and overall handling qualities seemed normal, although I didn't explore them in great detail. As Director of Flight Test, I was primarily interested in max performance and every time I flew it (I had three or four flights in it) I always took it out to its maximum performance. I would contribute then, my remarks to the pilot's ob's (observations) that went into the report. I tried to fly them all and get familiar with each project, because the pilot's report was going to have to go through me. So obviously, I tried to fly every airplane out here".

"Concerning program cancellation, General Boyd told me that the cost was just too high for the program and that the Air Force could not afford to buy-in at that cost. The alternative was the B-57, or the Canberra which we bought from the British. We bought all the engineering drawings, all the rights, and when you costed that total program out, it came out much less than the cost figures for the XB-51. For instance, the research and development were already paid for. So, the dominant-dominant-reason for not buying the XB-51 was cost of the total program".

On 8 February 1952, Lieutenant Colonel Robert L. Cardenas received his check-out in the XB-51. Now a retired Brigadier General, Cardenas is best known for being the Chief Test Pilot of the YB-49 flying wing bomber program. "I was Chief of the Bomber Section, and as a Bomber Chief my boss, Colonel Boyd, expected me to fly every bomber in the test. So when the XB-51 came along I was quite eager to fly it".

"As I remember it, it was extremely easy to fly. We were concerned with cross-wind landings because of the bicycle landing gear. But, when you came in for a landing and you put your flaps down the incidence automatically adjusted, actually bringing the nose down, and it slowed you down to about 90 knots. Approach was very flat and you touched down on both forward and aft gear simultaneously. The new Decelastat brakes further aided in a very short landing roll. I remember B-57 short field landing tests. The Canberra landed short, rolling past his mark blowing tires trying to stop. I was amazed when they didn't buy the '51 and turned around and bought the Canberra".

"Regarding the top speed of the airplane, I was told that it flew about 50 knots faster than the design and the powerplants should have allowed it. All of the test data said that it was flying 50 knots faster than it was designed to. They thought it was the engines. So they took the engines out and sent them to calibration. The engines calibrated within spec's. So then they thought that with the engines installed, combined with the design on the inlet ducts, it might be causing some additional thrust. So they built a trim pad out on the dry lake bed edge at Edwards. They dug a huge hole and filled it with cement and anchored it (the XB-51) onto the lake. Next they conducted a series of engine runs hoping to learn something. They didn't discover any additional sources of thrust during these tests. The thrust was as designed".

"Additional flights were made at various altitudes and settings, and it came out that it just flew 50 knots faster. Since then it has been suggested that this may have been an early example of Richard Whitcombs Area-Rule principal in action".

One of the pilots with the most time in the XB-51 is retired Air Force Brigadier General Charles "Chuck" Yeager, the first man to break the sound barrier. Yeager's other accomplishments include high-speed, and high speed weapons delivery testing on the XB-51. On one occasion he recalls laying 4,000 pound mines at .92 MACH in close proximity to the ground and the water at El Centro, California. Yeager remembers: "I flew both ships 1 and 2 and never had a hairy moment in the airplane". He also remembers SHORAN bombing tests, when he flew five flights, which included loading bombs, taxi tests,

Above, the first man to break the "Sound Barrier" Major Chuck Yeager before a flight in the XB-51. Crew chief Earnie Philson is at Yeager's right. Below, Major Robert Cardenas, who piloted the B-29 mothership for Yeager's X-1 historic flight, also flew the XB-51 and became a Brigad'ereGeneral like Yeager. He poses here in front of a German Arado AR-234 jet bomber. (USAF via Libis)

At left, close-up of the Air Force Flight Test Center badge painted on the side of 46-685. (via Wayne Morris)

take-off, flight to range, bomb run and return, in 2.5 hours total time.

"In a high speed dive the plane would max-out at .93/.94 Mach. The airplane was way ahead of its day. One problem was that Glenn L. Martin sided with the Navy in the B-36 flap". Yeager felt that this may have been a contributing reason for Martins failure to win a production contract. Regarding the untimely demise of both ships 1 and 2, Yeager offers, "both aircraft were destroyed by 'bomber pilots'. Too bad, they were very good machines".

Other notable test pilots checked out in the XB-51 were Lieutenant Colonel Fitzhugh Fulton, Colonel Russ Schleeh, Brigadier General Guy Townsend, and Major General Albert Boyd.

CONCLUSION

Canceled in November 1951, the program was not officially closed-out until October 1953. The final payment to Martin covered repairs for a previous landing accident, the modifications of ship #1 in preparation for the bomb dropping tests. Over a-four-year period, the total development costs went from 9.5 million to 12.6 million dollars.

The fact that both of the XB-51s were lost in accidents cannot be attributed to aircraft performance. Causes of both accidents were clearly pilot error and in no way suggest a design problem with the airplane. Landing mishaps experienced by the airplane were not considered excessive considering the learning curve for new airplanes at that time.

Selection of the B-57 Canberra may just have been that the range and loiter capabilities of the Canberra outweighed the speed advantage of the XB-51. In addition to this, the Canberra was already in production, and the Air Force needed to field the A-26 replacement as quickly as possible for the Korean conflict. Although arriving too late for Korea, the Canberra went on to see service in Viet Nam and proved to be a fine airplane. Without a surviving example, the XB-51 has faded into obscurity. Fortunately, the Air Force and the aviation industry recognized and capitalized on the XB-51's strong points. This program yielded much useful data in the areas of bicycle-type landing gear, variable incidence wing, advanced tail configuration and the rotary bomb bay door. This design data was invaluable to subsequent aircraft designs such as the B-47, B-52, B-57, F-8 and many others.

Not to be realized, was the airplane's potential for testing new engine designs. The #2 engine bay was spacious enough to accommodate a larger powerplant with afterburner. Located on the aircraft's centerline, the #2 position was ideal, as it would not have caused asymmetrical thrust conditions during testing. Engines #1 and #3 could have been retained in their original configuration for an added margin of test safety. The XB-51 could then have provided valuable inflight performance data, while serving as a testbed for new engine technology. It is quite possible that a simple engine upgrade, such as this, would have enabled the XB-51 to break Mach 1, making it the world's first supersonic bomber.

Martin's final aircraft design was another jet, a patrol bomber, the P6M Seamaster. This airplane would attain Mach 1 and go on to become Martin's crowning technical achievement in flying boats. But that is another story.

At left, Martin's last gasp, the superlative P6M Seamaster, shows its XB-51 heritage in the use of the "T" tail and fillet. The P6M was capable of a sustained low level dash speed of .89 Mach. (Martin)

MARTIN XB-51 "PANTHER" MODELS

HOBBY-TIME:
The XB-51 was tentatively named "Panther" even though it was never officially named. Hobby-Time offered a balsa wood and plastic kit which had to be carved and shaped from blocks of balsa using the instructions provided.

EXECUFORM:
One of the 64 kits of unusual vacuformed US aircraft offered by Execu-form is the XB-51. It comes with four sheets of drawings and white metal detail parts. Because of time constraints we were unable to build an example for the book. In 1/72 scale its the only kit available and can be purchased by contacting Mike Herrill at P.O. Box 347, Cedar Glen, CA 92321. Sample curtesy of Mike Herrill.

HBM 1/200 RESIN MODELS:
Prof. Ron Crawford, P.O. Box 23, N. Ferrisburgh, VT 05743, has a product line of about 520 types which include the XB-51. The model below was built by Scott Libis who had custom decals made for the National Insignias, tail numbers and the Martin XB-51 nose markings. The decals were made by Fred's Graphics, fhallstein@aol.com. The model was built with the early "T" tail.

1/48 SCALE COLLECT - AIRE MARTIN XB-51

Martin XB-51

A Magnificently Produced Model of this Trend Setting, But Little Known Attack Bomber from the 1950's!!!

Kit features; Huge finely detailed resin castings with fully engraved detail! Full Cockpit detail with all bulkheads! Crew hatch and ladder! Intricate, highly detailed landing gear! Numerous Metal parts! Fully detailed Rotary Bomb Bay! Optional parts for all variants! Full Engine, Exhaust, Wheel well detail and Depth! Clear Vac. Engraved Canopy! Custom Colorful Decals!

Another Model that you NEVER thought you would see!

MID-TECH SERIES #4833

A CUSTOM 1/48 Scale LIMITED PRODUCTION KIT!

Collect-Aire Models, Inc.

Big and impressive describes the Collect-Aire 1/48 scale XB-51. It is a mixed media kit made up primarily of resin castings supplemented by white metal detail parts, a vacuform canopy and colorful decals for all schemes carried on both prototypes. It is a limited issue kit with only a few hundred produced. These kits are available directly from Collect-Aire Models, 166 Granville Lane, North Andover, MA 01845. Or call (978) 688-7283, or FAX (978) 685-0220. All the kits detail parts were used in the model, including a full bomb load. The flaps and slats were extended, yellow JATO bottles were installed as was the crews entrance hatch and ladder. Photos and information curtesy of Collect-Aire Models.

PAGE 49 TOP TO BOTTOM: The XB-51 was unofficially called the "Panther". Scott Libis illustrates one of the logos the name inspired. Tandem nose proposal was labeled "Tactical Bomber" by Martin artists. Martins propensity for Navy flying boats is evident in the last three designs. The length of the XB-51s fuselage made for an easily blended hull. The conventional hull design was followed by a hydro-ski version and then a hydro-sled version. Remounting the podded forward engines atop the fuselage made these design concepts possible. (Martin via Stan Piet)